NATIONAL SECURITY IMPLICATIONS OF GLOBAL WARMING POLICY

> Claims that climate change is accelerating are bizarre. There is general support for the assertion that GATA has increased about 1.5° Fahrenheit since the middle of the 19th century. The quality of the data is poor, though, and because the changes are small, it is easy to nudge such data a few tenths of a degree.
>
> —Richard S. Lindzen,
> Massachusetts Institute of Technology
> Professor of Meteorology

Although numerous historical examples demonstrate how actual climate change has contributed to the rise and fall of powers, global warming, in and of itself, is not our nation's greatest climate threat. Past climate change episodes have simultaneously produced favorable and unfavorable conditions to which some nations successfully adapted while others failed. While significant environmental changes could certainly impact both America's standard of living and national security, the nation has the size, variety, and technology necessary to adapt to new climatic conditions. Today, the greatest climate threat to our national security is not actual change, but rather the world's perception of climate change and the resulting governmental and intergovernmental policies enacted to reduce the theorized anthropogenic greenhouse warming. As the world's governments become convinced that global warming is universally bad and that humans are the primary cause, political leaders may develop ill-advised policies restricting the United States' access and use of global energy supplies, weaken the United States' economy, and unfairly advantage rising developing nations. The United States' energy, economic, and security policies are inextricably linked, and poorly conceived multilateral policies could combine to threaten United States national security by reducing our relative national power in comparison with developing nations such as Brazil, Russia, India, China (BRIC countries). In the interest of national security, the United States should reject the multilateral approach championed by the United Nations Intergovernmental Panel for Climate Change and pursue unilateral initiatives focused on national self interests. Rather than adopting multilateral policies aimed at reducing the carbon emissions of developed nations, the United States should continue to resist adopting Kyoto Protocol type policies to preserve our national wealth to better fund Homeland Defense and national security.

Historical Climate Change Impacts: The Vikings as a Case Study

Today's common, populist view of global warming is generally catastrophic and includes a deep-seated belief that humans cause global warming through the intensive burning of fossil fuels. Anthropogenic global warming advocates often paint a dark picture of devastating droughts and crop failures, melting Arctic and Antarctic ice caps, rising sea levels and inundated coastlines, more severe hurricanes and other storms, and highly destructive epidemics. The most extreme advocates even predict extinction of the human race if anthropogenic global warming isn't immediately addressed and reversed.[1] Fortunately, recent human history indicates otherwise. Although there are numerous historical examples, worldwide, where climate change has apparently influenced the rise and fall of nations, the Vikings provides perhaps the best case of a rise and fall in power influenced by climatic change since their power swing closely correlates to well-documented warming and cooling climates. The Viking rise to power corresponded with the warmest four centuries in Europe for the previous 8,000 years called the Medieval Warm Period (800-1300 A.D.).[2] Subsequently, as the northern European climate again turned cooler, around 1200 A.D., the Norse Empire began a slow decline ultimately coinciding with the Little Ice Age, beginning around 1300.[3]

In a nutshell, warming in northern Europe allowed the Norsemen to develop more varied agriculture, experience higher crop yields, and produce abundant timber giving them an economic advantage over their surrounding neighbors; the increased agricultural productivity resulted in population growth which provided the Vikings with an additional advantage - manpower for soldiers and sailors. For several centuries, the Vikings turned these economic and population advantages into military strength. Their military power, centered on longships crewed largely by adventurers and part-time raiders, was a direct benefit of higher agricultural productivity. They used the military to raid and plunder tremendous swaths of territory including the British Isles, the Frankish Empire, and Russia. Additionally, they established colonies in Iceland, Greenland, and eastern North America. Important to the Viking expansion was their increased ability to harvest agricultural crops and natural resources, primarily wood and iron. Between AD 800-1000, the considerably warmer climate allowed both crops and forests to grow 100-200 meters higher in elevation than had been the historical norm thus increasing the acreage for both farming and forestry.[4] The thriving forests provided plentiful wood supplies which made plank-work cheap and helped give rise to the era of Viking longships.[5] In addition to longships, they built a commercial fleet of fishing and cargo vessels to extend their economic advantages through trade.

During their era of expansion, the Vikings discovered Iceland in the mid 9th century and settled the island in 874; by around 930, over 20,000 settlers called Iceland home.[6] Between 982-985, Eric used Iceland as a base of operations to discover and colonize Greenland.[7] The Vikings established two major settlements near modern-day Julianehaab and Godthaab and summer seal hunting camps in Baffin Bay. Early on, climate favored the colonists with temperatures warmer than today, as evidenced by the abundance of cod and the ability to farm.[8] The colonies prospered and attained a population of around 3,000 farmers, fisherman, and hunters.[9] However, in the early 1300s, the colonies began to suffer as the climate cooled. By 1342 the traditional Iceland-Greenland sailing route had to be shifted farther south due to ice, and in 1369, regular ship trade between Greenland and Norway ceased. Following prolonged multi-decadal cooling, an unusually cold period between 1350 and 1380 rapidly accelerated the decline of the colonies.[10] The increasing cold destroyed summer crops, limited access to the northern hunting grounds, and caused fish to migrate southward away from the Greenland coast. To add to the colonist's misfortunes, in 1349 the Black Death entered Norway through the port of Bergen and spread to both Greenland and Iceland. Although population records are not precise, the mortality rates were 45-55% in Scandinavia proper, 60% in Iceland, and potentially 100% in Greenland. The high mortality rates, essentially double the rates of the rest of Europe, were attributed to the colder climate which facilitated pulmonary complications leading to pneumonic plague – in addition to bubonic plague.[11] The devastating plague was the final nail in Greenland's coffin, and the colony never recovered.

Ultimately, the Greenland colonization failed for two reasons: the gradually deteriorating climate and the southern expansion of Eskimos which was also cold related.[12] By 1350, the more northerly settlements on the west coast were uninhabitable due to a combination of the increasingly cold climate and attacks by the Inuit Eskimos. Further south, agriculture became impossible, with repeated crop failures and loss of livestock producing starvation conditions. Finally, Norway steadily declined as a seapower during the 15th century and lost its ability to resupply the Greenland colonies.[13] Around 1500, the final colonists succumbed, and the Viking Greenland saga ended.[14] While the Inuits continued to live in isolated settlements, Europeans did not return to Greenland until around 1721 followed by a Danish territorial claim in 1775. Today, the country's settlement pattern closely resembles the old Viking colony.

Elsewhere, the Viking Empire continued to decline in concert with the expanding Little Ice Age. Worldwide, Viking colonies steadily declined and were either overtaken or assimilated into the local cultures they had once dominated. The Scandinavian nations comprising the Viking Empire degenerated into individual nation states, notably Norway. By the 16th century, all three major components of the Norwegian economy – fishing, agriculture, and forestry – were negatively affected by the increasing cold. Colder seas caused the once fertile fish stocks of cod and herring to migrate southward, a boon for Denmark and England; agricultural crop yields declined due to shorter growing seasons; and the colder climate concentrated the timber industry in the south.[15] While not entirely due to climate change, political and military power shifted southward to the emerging European world powers of England, Spain, and France.

Interestingly, the northern European experience during the Medieval Warm Period and Little Ice Ages were not globally uniform events with constant effects. In the Americas, this same episode of warming is closely linked to severe, multi-decadal droughts and widespread crop failures possibly leading to the demise of the Mayan Empire shortly after 900 A.D.[16] With the collapse of the Mayan civilization, the Americas were left with no great power until the Europeans arrived in the 17th century.

In eastern Asia, the Khmer Kingdom of Cambodia reached the pinnacle of its prosperity between 900 and 1200. The king had built a complex series of canals to move water throughout the kingdom which solidified his central power and authority. The warming climate apparently created ideal monsoons for the kingdom's centralized approach to governing and their level of technological advancement.[17] However, beginning with the onset of the Little Ice Age, the cooling climate upset the favorable monsoon pattern and the drought stricken empire rapidly collapsed as its intricate canal system failed. Lastly, in China, the warm period between 900-1200 created drought conditions across much of the country. The resulting competition for resources split the three century old T'ang Dynasty into fifteen fragmented kingdoms.[18] Interestingly, the rise of the Ming Dynasty (1368-1644) closely corresponds with the start of the Little Ice Age when regionally improved climatic conditions restored favorable monsoon conditions to much of China. In summary, regional climate change had a positive effect on Viking expansion, a negative effect on European powers attacked by the Vikings, a devastating effect on the only empire in the Americas, and mixed effects throughout Asia.

In terms of climate, the Viking story offers numerous lessons. First, the climate can dramatically warm and cool due to natural variations; additionally, these natural climatic changes have occurred during relatively recent human history. While ardent anthropogenic induced warming theorists often dismiss the Medieval Warm Period, the proxy data outlined by H.H. Lamb in Climate, History and the Modern World, clearly supports its occurrence; additionally, Lamb's climate proxy research indicated the Medieval Warm Period was warmer than our contemporary climate. Second, the terms global warming and global cooling are misnomers because climate change is a regional phenomenon, with highly variable effects, versus a global event. Looking at climate change through a global lens provides an inaccurate view as demonstrated by the different experiences of Scandinavia, China, and North America during the same warm period. Third, climate change, regardless of temperature and precipitation direction, has the greatest effect on the availability of and access to the natural resources that form the basis of national economies. During the Medieval Warm Period, the Scandinavian Vikings prospered due to favorable conditions for agriculture, forestry, and fishing while the Mayans suffered devastating droughts that destroyed their crops. Fourth, and most important, this regional, natural climate change helps produce clear winners and losers amongst the world's nations. The Vikings, as did the Chinese, proved to be both winners and losers during different phases of climate change. The primary lesson of history is that climate change helps produce clear economic winners and losers.

While much can be learned from studying the past, today's world is not the same as that of the Vikings, Mayans, and Khmer. The population difference is striking. In 1000 A.D. the UN estimates there were 310 million people, worldwide, while today's UN population estimate exceeds 6.5 billion. This population increase, combined with technology and governance, has contributed to a highly globalized economy where changes in one market often ripple throughout the world. Due to the interrelated nature of modern economies, an individual nation's economic decisions can help or hinder numerous other nations. A single economic decision, today, could be far more devastating to a country than a fleet of Viking Longships.

Global Warming Theories

Although many politicians strongly espouse that anthropogenic global warming is a fact of settled science based on IPCC reporting, few climate scientists, notably

Massachusetts Institute of Technology Professor Dr. Richard Lindzen, agree. In fact, most climate scientists believe climate changes in response to numerous, combined factors.[19]

> The notion that complex climate "catastrophes" are simply a matter of the response of a single number, GATA, to a single forcing, CO_2 (or solar forcing for that matter), represents a gigantic step backward in the science of climate.[20]

The climate system is a complex, interactive system consisting of the atmosphere, land surface, snow and ice, oceans and other bodies of water, and living things. The atmospheric component of the climate system most obviously characterizes climate. While climate is commonly thought of as the average weather, a more precise description is the mean and variability of temperature, precipitation, and wind over a period of time, ranging from months to millions of years (the classical period is 30 years). The climate system evolves in time under the influence of its own internal dynamics and due to changes in external factors, or forcings, which affect climate.[21]

There is no single theory of climate change but rather competing emphases on various components of the overall climate system depicted in Figure 1, below. The United Nations IPCC has chosen to place its primary emphasis on anthropogenic emission of greenhouse gases, primarily carbon dioxide.[22] In their numerous reports on climate, the IPCC acknowledges the sun is the primary driver of climate through solar radiation but insists the observed temperature increase since 1850 is primarily caused by human influences. Other scientists, of which there are many, believe the sun is the dominant influence and greenhouse gases play a much smaller role.

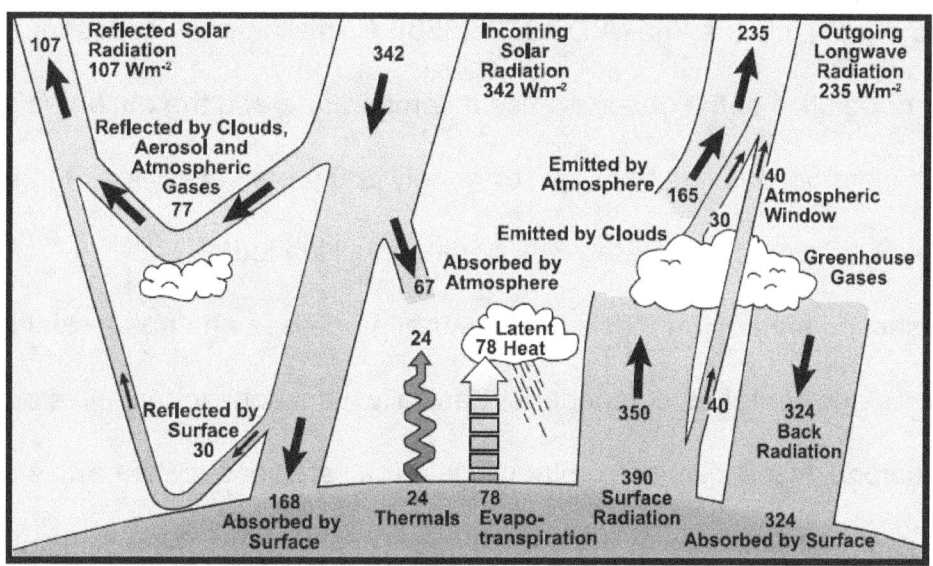

Figure 1. Solar radiation powers the climate system. There are three fundamental ways to change the radiation balance of the Earth: 1) by changing the incoming solar radiation (e.g., by changes in Earth's orbit or in the Sun itself); 2) by changing the fraction of solar radiation that is reflected (called 'albedo'; e.g., by changes in cloud cover, atmospheric particles or vegetation); and 3) by altering the long wave radiation from Earth back towards space (e.g., by changing greenhouse gas concentrations).[23]

The anthropogenic greenhouse advocates emphasize that atmospheric gases (such as carbon dioxide, methane, and water vapor) form a blanket for the earth's atmosphere which captures solar heat and subsequently transfers that heat to the world's oceans. This heat engine gradually causes the earth's temperature to rise and thereby influences the climate. An essential part of this theory is that humans are responsible for releasing greater than natural amounts of greenhouse gases through our use of hydrocarbon fuels, intensive farming, and extensive deforestation.[24] While sounding relatively simple, this is a complicated theory involving complex energy transfers between the atmosphere and the hydrosphere. Proponents of this theory tend to use a timeline beginning around 1850 to illustrate a warming of 0.6° Celsius, or about 1.5° Fahrenheit, and link the warming to anthropogenic influences beginning in the industrial revolution; this timeline start date is not insignificant as it also corresponds with the end of the Little Ice Age (A.D 1350-1850). A key line of evidence for this theory is the close correlation of the observed rise in temperature and the observed global average rise in carbon dioxide from approximately 280 parts per million before the industrial revolution to approximately 385 parts per million today.

The second area of emphasis is on natural processes and has a significant heliocentric, or sun centered, component. Basically, variations in the sun's output (sunspots, coronal mass ejections, solar winds, etc), variations in the earth's orbit around the sun, and variations in the earth's movement through the ecliptic plane combine to dictate the amount of solar radiation reaching the earth's surface. Increased amounts of solar insolation result in global warming while reduced insolation results in net cooling. As with the anthropogenic greenhouse theory, this theory is also complicated as it also incorporates numerous factors including the greenhouse effect, regional changes in the earth's albedo, reflection by atmospheric aerosols and clouds, and regional oscillations such as the Pacific Decadal Oscillation, the Atlantic Multidecadal Oscillation and the El Nino/La Nina Cycle. Proponents of this theory point to a number of short, medium, and long term solar cycles. Key lines of evidence for this theory include the Maunder Minimum during the Little Ice Age, a long term periodic cycle called the Milankovic Cycle, and climatic variations caused by the oscillations.

Anthropogenic global warming advocates generally select 1850 as the base year from which to illustrate man-made warming. Ostensibly, 1850 was selected due to its close correlation with the Industrial Revolution and the resulting increase in human produced greenhouse gas emissions. However, from a climatic viewpoint, 1850 was effectively the end of the Little Ice Age. The series of diagrams, below, indicate 1850 was one of the coldest periods in modern human history which may lead a skeptic to the conclusion that warming would naturally occur after an ice age.

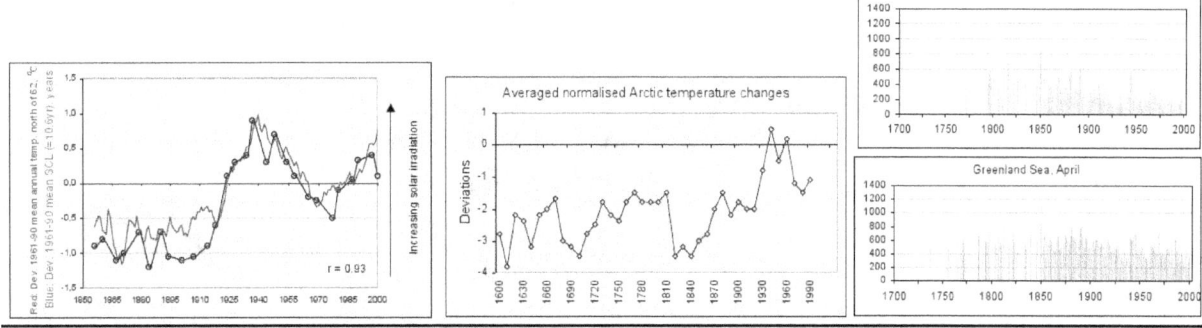

Figure 2. These three charts provide measurements indicating 1850 was one of the coldest years in modern human history. The chart on the left compares solar cycle lengths and temperature; the middle chart provides averaged Arctic temperatures, and the chart on the right illustrates Greenland and Iceland sea ice maximum.[25]

Two questions must be asked of the IPCC. First, if European temperatures during the height of the Medieval Warm Period, a span of about 150 years, was 1°-1.4° C warmer than today's temperature[26] without any comparable anthropogenic release of carbon dioxide, then why have human emissions only caused a rise of 0.6° C over a similar timeframe. Second, if atmospheric carbon dioxide levels are currently higher than at any time in modern human history, why has there been no statistically significant warming since 1995?[27] Despite the Panel's efforts since 1988 to establish a consensus that humans cause global warming, the only consensus is that human induced carbon dioxide is one of many components driving climate. A good example is a recent research article published in the journal, *Science*, attributing at least 30% of the surface warming in the 1990s to higher stratospheric water vapor levels; subsequently, the study indicates a 10% decline in water vapor levels accounts for the essentially flat temperature increase trend since 2000.[28] This type research clearly shows the science is far too unsettled and uncertain to definitively assert that eliminating carbon emissions will halt climate change. As former National Aeronautics and Space Administration climate scientist and University of Alabama, Huntsville professor Dr. Roy Spencer says, "Climate change — it happens, with or without our help."[29]

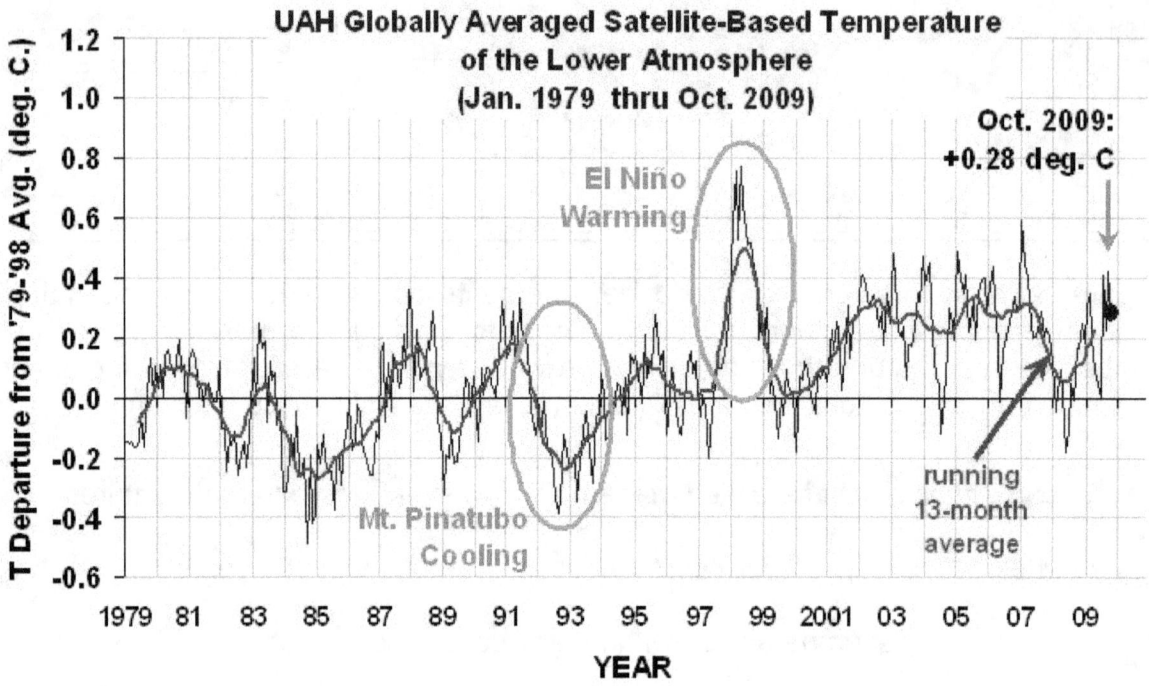

Figure 2. This chart depicts the flat temperature trend referred to in the above Science research article. Source: University of Alabama, Huntsville, Dr. Roy Spencer.[30]

The idea that humans cause global warming appeals to politicians and leaders because if people cause it, leaders can devise a plan to stop it. Conversely, asserting global climate change, whether warming or cooling, is primarily natural, and thus unstoppable, is a scary proposition. From the 1980s through 2009, the IPCC accumulated a considerable amount of evidence indicating anthropogenic driven increases in greenhouse gases were having a clear warming effect, globally. Within the last six months, however, the evidence has weakened. In November, scientists at the East Anglia University Climate Research Unit (CRU) and numerous prominent IPCC scientists became embroiled in a controversy deemed Climategate. Following an unattributed cyber attack on the CRU's network, numerous emails and documents revealed the organization had withheld evidence countering their scientific point of view, failed to maintain research data so their research could be peer reviewed, disregarded United Kingdom freedom of information release laws, and strove to discredit and limit peer reviewed articles not supporting the theory of anthropogenic global warming. Given the recent Climategate, new peer reviewed scientific publications supporting natural climate change processes, and the IPCC director's resignation due to inaccuracies in the latest report, there is insufficient evidence of anthropogenic global warming on which to base decisions impacting national security. If the emerging minority view that climate change is largely driven by natural variations and processes proves to be accurate, governmental action will neither be required nor effective in altering global temperature increases.

Governmental Response to Stop the Perceived Problem of Global Warming

Numerous countries have long sought to use the United Nations construct to either advance the anthropogenic greenhouse warming theory or to benefit from multilateral treaties. Other nations, including the United States, have often used the UN requirement for consensus as a means to contain international action. The Reagan administration initially proposed the Intergovernmental Panel for Climate Change as a replacement for self-appointed committees of scientists the administration thought were too alarmist.[31] When established in 1988, the IPCC was designed to require unanimous international consent for any official statements and thus both greenhouse theorists and conservative governments both saw the body as a good compromise.

Over time, the Reagan administration's hope for a conservative, stalemated organization failed to materialize. Instead, the IPCC was instrumental in establishing two milestone treaty documents. The 1991 United Nations Framework Convention was ratified by 198 countries, including the United States, and established the legal groundwork placing the responsibility for greenhouse gas reduction on developed nations. The 1997 Kyoto Protocol built upon the 1991 Framework by instituting legally binding emissions reductions for industrialized nations and introduced the concept of emissions trading. The Kyoto Protocol was ratified by 169 countries with the United States and Australia notably refusing to adopt the treaty.[32] These major international agreements were the products of tremendous amounts of research, negotiation, and consensus by IPCC scientists and policy makers which was published in IPCC Assessment Reports. More recently, the IPCC championed the authoring of an additional binding document at the Copenhagen meeting in December 2009; however, no binding agreement materialized due to a lack of consensus.

The European Union (EU) developed the European Union Greenhouse Gas Emission Trading Scheme (ETS) to implement the Kyoto Protocol. In 2003 the ETS entered into force as the world's first international carbon trading system. It is a cap-and-trade system based primarily on the free allocation of a fixed amount of emission allowances to a set of covered installations. Companies can either use these allowances to cover the emissions resulting from their production or sell them to other companies who need additional allowances.[33] The EU intends the ETS to be their centerpiece for complying with the Kyoto Protocol by regulating energy intensive installations including combustion plants, oil refineries, coke ovens, iron and steel plants, and factories making cement, glass, lime, brick, ceramics, pulp and paper.[34] Through implementing and operating the ETS, the EU plans to reduce their carbon emissions by 20% below the 1990 levels by the year 2020. ETS implementation consists of a series of phases in which each successive phase is more restrictive than the previous until the EU reaches its carbon reduction goals. The phases are broken down into multiple year trading periods and each country is allotted specific allowances with which to initiate the trading (see chart, below). The first trading period was from 2005 to 2007, the second period was 2008 to 2012, and the third trading period will begin in 2013. During the first trading period, all the credits were issued for free, a practice that diminished in the second trading period and may eventually end altogether.

Member State	CO2 allowances in millions of tonnes	Share in EU allowances	Installations covered	Kyoto target
Austria	99.0	1.5 %	205	-13%
Belgium	188.8	2.9 %	363	-7.5%
Denmark	100.5	1.5 %	378	-21%
France	469.5	7.1 %	1,172	0%
Germany	1,497.0	22.8 %	1,849	-21%
Greece	223.2	3.4 %	141	+25%
Ireland	67.0	1.0 %	143	+13%
Italy	697.5	10.6 %	1,240	-6.5%
Latvia	13.7	0.2 %	95	-8%
Luxembourg	10.07	0.2 %	19	-28%
Netherlands	285.9	4.3 %	333	-6%
Poland	717.3	10.9 %	1,166	-6%
Portugal	114.5	1.7 %	239	+27%
Spain	523.3	8.0 %	819	+15%
Sweden	68.7	1.1 %	499	+4%
United King	736.0	11.2 %	1,078	-12.5%

Table 1. This table is a sample of EU nations; the full list is available at the ETS official website.

Thus far there have been over $300 billion worth of carbon transactions.[35] As the EU continues to issue fewer and fewer free credits, both the individual credit prices and the total market prices are expected to increase dramatically. These transactions effectively increase the direct cost of generating electricity and the production of steel, chemicals, and cement. However, the credit trades are only part of the increased expense of the program. Whole government bureaucracies have developed to monitor and measure carbon emissions or to contract that work to private companies. These considerable expenses have not produced a measurable reduction in greenhouse gases. As an example, with carbon credit futures costing approximately $19.50 per ton (Feb 2010 European Climate Exchange), a megawatt of electricity produced by a coal fired plant would cost an additional $31. While this may sound like a small price to pay, in 2007, the US Department of Defense would have paid an additional $919 million based on using 29,656,103 megawatts of electricity.[36] From an Army perspective, that's essentially the personnel costs for three Brigade Combat Teams.

The United States Senate did not ratify the Kyoto Protocol; however, we are potentially on the road to cap and trade legislation similar to the EU ETC. In 2009, the United States House of Representatives passed the American Clean Energy and Security Act, based on the bill introduced by Representatives Waxman and Markey. The Environmental Protection Agency estimated the bill would result in $1.4 trillion a year going abroad to cover the offsets delineated in the legislation.[37] While the overall cost of the bill would be far higher, the offsets represent a complete drain on our economy as they would transfer money directly to foreign competitors, including China, who is already diversifying out of coal fired power plants for reasons not connected with climate change. The United States would, in effect, be paying for developing countries to build "green friendly" power plants they were already intending to build.....as Representative Bob Corker said, "That's not a market. That's Alice in Wonderland.[38]

In addition to affecting power plants, the House bill would also increase costs on facilities such as petroleum refineries. These additional costs would severely harm a relatively fragile industry. In January 2010, Chevron announced plans to cut jobs in its refining business and is considering completely exiting some markets. Conoco Phillips and ExxonMobil are both facing the same issues. Also, Valero Energy, a pure refiner, has recently closed a major refinery in Delaware, one of the largest on the East Coast.[39] While these layoffs and closures are primarily a result of the recent economic down turn, they highlight the paramount reason refining operations are increasingly unprofitable, the razor thin profit margins inherent to the business. Additional operating costs, like those created by the Waxman-Markey bill, would most likely drive refineries out of the United States and into developing nations. Businesses who choose to move offshore would no longer pay local, state, and federal taxes nor employ domestic workers. In addition to a diminished tax base to support national defense, the Department of Defense could end up depending predominantly on foreign sources for refined petroleum products such as fuels and lubricants.

Perhaps the most economically threatening component of either the EU ETS or the House cap and trade bill is the offset concept. Offsets find their origin in the Kyoto Protocol, and, unlike government issued carbon credits, offsets are a way for governments and private companies to develop new carbon credits based on approved projects. To monitor and sanction these offset credit earning projects, the Kyoto Protocol established a special United Nations organization called the Clean

Development Mechanism (CDM). The CDM, in turn, authorizes several private companies to review, evaluate, and substantiate proposed projects. The result is the United Nations effectively gains control of the world-wide supply of carbon credits in the form of a new commodity type security; however, these carbon securities are unlike any other commodity. Iron, coal, grain, crude oil, and other commodities are bought on the basis of a real, underlying asset that must be delivered on an actual future date. Carbon securities, however, are based on the non-delivery of an invisible substance that can be derived from both natural and manmade processes.[40] In short, carbon credits are intangible assets whose value is derived only from a United Nations treaty of which the United States is not a signatory.

The method through which the CDM approves carbon offset projects has significant potential for overestimating the amount of carbon saved while simultaneously overstating the emissions reduction. Given the United Nations' history with corruption, fraud, and the recent Iraq "oil for food" scandal, handing the organization a $3 trillion carbon credit market, based entirely upon the evaluation of carbon emissions reduction potential, seems unnecessarily risky. A recently approved Brazilian carbon offset project provides a perfect example of how overstating potential benefits could lead to "oil for food" style UN CDM fraud.

In 2001, Plantar, one of Brazil's largest forest resource companies, applied to the CDM for an offset program. In a nutshell, Plantar's plan was to convert 57,000 acres of overgrazed grassland savannah acreage to a eucalyptus tree plantation. They would then burn the eucalyptus trees in specially developed kilns that operated at low temperature and oxygen levels to create charcoal. The charcoal was then to be used in a nearby factory to produce pig iron ingots for sale to auto and appliance manufacturers. Along the way, this complicated process was verified by three different Designated Operational Entities (DOE) – companies certified by the UN to validate and verify proposed projects. SGS certified every ton of pig iron produced with charcoal instead of coal averted two tons of carbon dioxide. DNV validated the specialized kilns did, in fact, reduce methane emissions. And, TUV SUD verified that eucalyptus plantation absorbed more atmospheric carbon dioxide than overgrazed pastureland which it displaced. Ultimately, the project was awarded 12.8 million carbon credits over its 28 year operating life. Plantar traded 1.5 million credits to the World Bank to obtain funding for the project, and in 2002 began producing pig iron as a CDM project. In addition to obtaining funding for the overall project, Plantar is in the process of selling forward over $100 million worth of the issued carbon credits to various European banks and industries. To summarize the problem, Plantar was growing eucalyptus trees for charcoal to produce pig iron long before they applied for the CDM project status, and, in April of 2009, the company idled the entire operation.[41]

This single example offers more questions than solutions. Before it rises to the the same level scandal as "oil for food", can the CDM program managers enact safeguards and mechanisms to prevent unscrupulous, rogue actors from corrupting a well intended UN program through fraud and misappropriation of carbon credits? How can the CDM, through its DOEs, truly provide accurate assessments of future greenhouse gas emissions reductions if their system validated the Plantar scheme? If the Waxman – Markey bill were passed, and US companies were forced to buy developing nation offsets from schemes such as Plantar's, how would the public react to paying higher prices for goods and services only to benefit dubious foreign companies? These are serious questions we could easily avoid answering, as a nation, by simply foregoing the ratification of Kyoto Protocol and resisting foreign pressure to pass a cap and trade bill such as the current House Clean Energy and Security Act.

Conclusion and Recommendations

First, climate change has naturally occurred in recent human history. The US government needs to adopt this philosophy and espouse it in all international forums. During the Medieval Warm Period, the world saw powers in Europe and Asia prosper from warming and suffer from cooling. Conversely, powers in North America and China suffered from warming and benefited from climatic cooling. Whether primarily influenced by natural or anthropogenic processes, climate change is both inevitable and unstoppable, as evidenced by the human experience from 800 to present. Rather than attempt to devise programs and institutions to reduce carbon emissions, we should focus efforts and funding on preparing for the effects of climate change over rolling 20 year periods. While climate change episodes could potentially threaten future US national security, exorbitantly expensive governmental policy is the more serious threat. The US can develop technology and infrastructure to adapt to environmental conditions driven by climate change scenarios, either warming or cooling, as long as our economy is not threatened through multilateral, Kyoto Protocol driven policy dictating the transfer of trillions of dollars to developing nations. Every dollar US businesses are forced to inject into a program to purchase carbon offsets, purchase carbon credits, or for regulatory compliance reduces our GDP and thus our ability to fund national defense.

Second, our national security is too important to acquiesce to multilateral initiatives championed by developing nations in United Nations forums. A common argument for the US adopting aggressive carbon emissions reductions is that we risk losing international respect, influence, and power by being one of the few holdouts, along with Australia. However, despite the wide adoption of the Kyoto Protocol by many of the world's nations, when the US Senate steadfastly refused to ratify the treaty, we experienced no serious diplomatic or political repercussions. With its myriad of competing interests, giving the UN control of $2-$3 trillion worth of carbon credits vests far too much power in a single international governmental body. While major decisions in the UN are only made by the consensus of the member nations, all too often individual organizations, or individuals, within the UN have escaped sufficient oversight to prevent counterproductive activities. Given the lack of serious international repercussions and the risk of trillions of dollars flowing to potentially unfriendly developing nations, the US Senate should remain steadfast in its denial of the Kyoto Protocol, or any other similarly restrictive multilateral treaty.

Third, we must have an accurate understanding of climate change issues and science. It is in our interest to thoroughly study and analyze future climate change scenarios to determine the potential global winners and losers and to support both international relations and future strategy. For the climate "winners", we must devise strategies to incorporate, prevent, or minimize their power; and, for the "losers" we must decide whether or not to support as a partner or ally. There is an old adage, "you usually find what you're looking for," which is particularly applicable to the climate problem in light of the recent Climategate scandal involving the East Anglia University Climate Research Unit stifling the publication of peer reviewed research articles. The US needs to emphasize the importance of natural processes in regard to climate change and rebalance our national scientific research funding to reflect that view point. While there is considerable funding and effort concentrating on predicting the future through climate modeling, those models rely on numerous assumptions and incomplete data. By distributing research funding to focus on climate and anthropological studies of the recent warming and cooling episodes during the Medieval Warm Period and the Little Ice Age, we could potentially better predict the future by more thoroughly understanding the past impacts on society. Additionally, the increased understanding of natural climate change processes would aid our international negotiations and provide a strong basis for refuting the need for international cap and trade schemes similar to the European Trading System.

To conclude, there is currently insufficient evidence of anthropogenic global warming on which to base policy decisions impacting national security. The emerging minority view emphasizes natural variations and processes as the primary drivers of climate change as opposed to anthropogenic greenhouse gases. If this view proves to be accurate, governmental action will neither be required nor effective in altering global temperature increases. Rather than adopting multilateral policies, programs and institutions aimed at reducing the carbon emissions of developed nations, the United States should continue to resist adopting Kyoto Protocol type policies to preserve our national wealth. While climate change episodes could potentially threaten future US national security, exorbitantly expensive governmental policy, with its unforeseen and unintended consequences, is the more serious, and immediate, threat.

Endnotes

[1] Robert Henson, The Rough Guide to Climate Change (New York: Rough Guides, Ltd, 2008), vii.

[2] Brian Fagan, The Little Ice Age: How Climate Made History: 1300-1850 (New York: Basic Books, 2002), 7.

[3] Ibid, 47.

[4] H.H. Lamb, Climate, History and the Modern World, 2d edition (London and New York: Rutledge, 1995), 177.

[5] T.D. Kendrick, A History of the Vikings (Mineola, NY: Dover Publications, Inc, 2004), 24.

[6] Ibid, 336.

[7] Ibid, 361-364.

[8] Lamb, Climate, History and the Modern World, 2d edition, 175.

[9] Kendrick, A History of the Vikings, 362-364.

[10] Fagan, The Little Ice Age: How Climate Made History: 1300-1850, 9.

[11] Robert S. Gottfried. The Black Death (New York: The Free Press, 1985), 57-58.

[12] Kendrick, A History of the Vikings, 368.

[13] Ibid, 365.

[14] Lamb, Climate, History and the Modern World, 2d edition, 187-188.

[15] Fagan, The Little Ice Age: How Climate Made History 116.

[16] Brian Fagan, The Great Warming (New York: Bloomsbury Press, 2008,141.

[17] Ibid, 207-8.

[18] Ibid, 222-224.

[19] Richard S. Lindzen, "The Climate Science Isn't Settled", *The Wall Street Journal*, 1 December 2009, A19.

[20] Ibid, A19.

[21] International Panel for Climate Change, "Historical Overview of Climate Change", *Climate Change 2007: The Physical Science Basis. Contribution of Working Group I to the Fourth Assessment Report of the Intergovernmental Panel on Climate Change* (Cambridge University Press, Cambridge, United Kingdom and New York, NY, USA), 96.

[22] Ibid,100.

[23] Ibid, 96.

[24] Henson, The Rough Guide to Climate Change, 36.

[25] Torgny Vinje and Hugues Goosse, "Ice Extent Variations in the Last Centuries", briefing poster, St Petersburg, Arctic and Antarctic Research Institute, 11-14 November 2003, 2-4.

[26] Lamb, Climate, History and the Modern World, 2d edition, 179.

[27] Phil Jones, BBC Interview, 13 Feb 2010, http://news.bbc.co.uk/2/hi/science/nature/8511670.stm

[28] Susan Solomon, Karen Rosenlof, Robert Portmann, John Daniel, Sean Davis, Todd Sanford, and Gian-Kasper Plattner, "Contributions of Stratospheric Water Vapor to Decadal Changes in the Rate of Global Warming", *Science*, 28 January 2010.

[29] Dr. Roy Spencer website, http://www.drroyspencer.com/

[30] Ibid.

[31] Spencer R. Weart, "Global Warming as a Scientific Puzzle", Global Climate Change: National Security Implications, (Carlisle Barracks, PA: The Strategic Studies Institute), 38.

[32] Dennis Tanzler, "International Diplomacy", Global Climate Change: National Security Implications, (Carlisle Barracks, PA: The Strategic Studies Institute), 188.

[33] International Panel for Climate Change, Climate Change 2007: The Physical Science Basis. Contribution of Working Group I to the Fourth Assessment Report of the Intergovernmental Panel on Climate Change, 96.

[34] European Union Emission Trading System official website, http://ec.europa.eu/environment/climat/emission/index_en.htm accessed 31 January 2010.

[35] Mark Shapiro, "Conning the Climate, Inside the carbon-trading shell game", *Harpers Magazine*, Feb 2010, 31

[36] Anthony Andrews, "Department of Defense Facilities Energy Conservation Policies and Spending" (Washington, D.C.: The Congressional Research Service, 19 Feb 2009),10.

[37] "Cap and Tirade", *The Economist*, 5 December 2009, special edition 14.

[38] Ibid, 15.

[39] Ann Davis and Isabel Ordonez, "Chevron to Cut Refining Jobs", *The Wall Street Journal*, 20 January 2010, B3.

[40] Shapiro, "Conning the Climate, Inside the carbon-trading shell game", 32.

[41] Ibid, page 35.

www.ingramcontent.com/pod-product-compliance
Lightning Source LLC
Chambersburg PA
CBHW081824170526
45167CB00008B/3532